Manfred G. Riedel

Winning with Numbers

A KID'S GUIDE TO STATISTICS

illustrated by Paul Coker, Jr.

PRENTICE-HALL, INC.
Englewood Cliffs, New Jersey

Printed in the United States of America   J

Prentice-Hall International, Inc., London
Prentice-Hall of Australia, Pty. Ltd., North Sydney
Prentice-Hall of Canada, Ltd., Toronto
Prentice-Hall of India Private Ltd., New Delhi
Prentice-Hall of Japan, Inc., Tokyo
Prentice-Hall of Southeast Asia Pte. Ltd., Singapore

10   9   8   7   6   5   4   3   2   1

Library of Congress Cataloging in Publication Data

Riedel, Manfred G., 1937-
  Winning with numbers.

  Summary: A simple explanation of statistics and
their application to everyday life.
  1. Statistics—Juvenile literature.
[1. Statistics] I. Coker, Paul. II. Title.
HA29.R56   001.4'22   77-26729
ISBN 0-13-961466-4

CONTENTS

INTRODUCTION

No matter what you do in life, you're probably going to run into statistics. Statistics is the science of collecting and classifying numerical facts, and of using these data to draw conclusions.

If you become a biologist, politician, or swimming coach you'll need statistics on tree rings, on election votes, or on competitive races. Many of us run into statistics without noticing it. In a tourist office or in school you may have been asked to fill out a simple form about how you liked the Grand Canyon, for example, or what television shows you watch. Even though no numbers were involved in your answers, you yourself were in the process of becoming a statistic! You may have unwittingly contributed to a huge survey which later will have impact on other Americans going to the Grand Canyon or watching television.

So important is this science of collecting facts and drawing conclusions that statistics is even mentioned in the U.S. Constitution. By law the government has to count the people and measure the land, it has to provide numerical information on the labor force, employment, wages, education, law enforcement, climate, transportation, and so on. And now, with the help of computers, this type of data can be collected faster and more easily than ever before in history.

Statistics appears in almost every human activity—from astronomy to zymurgy*! And H.G. Wells, who liked to think about the future (he wrote **The Time Machine** and **The Shape of Things to Come**) predicted that statistical

---

*According to Webster, zymurgy is the branch of chemistry dealing with fermentation, as applied in wine making, brewing, etc.

thinking will one day be as necessary as the ability to read. Already today, to buy a bicycle or a can of tomato juice, you—as an intelligent consumer—have to be able to compare products and prices. And understanding statistics is the best way to come out a winner.

But if this science is so important, why do so many people say they don't believe in statistics?

Why do people say that "you can prove anything with statistics"?

Why do people claim they always fall asleep reading statistics?

1 WHAT IS STATISTICS?

Modern science has sometimes developed from a curiosity that is centuries-old. Astronomy, now a complex study conducted with sophisticated equipment in observatories, probably started with early man's wonder about the moon and stars.

You think of numbers, and arithmetic comes to mind. That science very likely started with primitive people's need to keep track of their arrows or their children. What began with counting on fingers has today become a highly sophisticated body of ideas about numbers.

So, it seems, the history of statistics should also go back thousands of years. Yet it was not before the 17th century that Blaise Pascal, a French scientist and philospher, began the studies that led to statistics. And he did so because of gambling. In those days gambling was a common pastime, and Pascal used arithmetic to figure out what chance a player had of winning various games. These studies in the probability of winning games of chance eventually led to modern statistics.

Statistics had still another, very different, origin. In the 18th century, politicians began to discover that statistical methods could help them learn a great deal about voters and thereby enable them to win election.

So through gamblers and politicians, both interested in increasing their chances of winning, statistics developed. And with time, the term statistics has acquired more than one meaning:

Statistics **are** all those numerical facts (or data) that you find in the sports or financial pages of a newspaper or in yearly almanacs.

Statistics **is** the modern science (or theory) dealing with the gathering of facts, with taking samples, and—of ever increasing importance—with using these data to make inferences and predictions.

If **you** like to gamble, try answering the questions at the end of each chapter. You'll find the answers at the end of the book.

1 **Question:** Why are some gamblers surprised to lose?

2 GATHERING FACTS

Takeoff

"What are you doing?" Sheila asked Burt when she found him climbing up a tree trunk.

"Counting owl eggs," said Burt, pointing up to a hole in the tree. "Owls swallow mice whole, digest them, and then cough up the bones and fur in a pellet near the nest or roost. So by these pellets on the ground I know there's an owl up there, but I have to wear a helmet because some owls divebomb you when you get near the nest."

"You're kidding! Counting eggs—how boring can you get!" said Sheila.

"Somebody has to do it," said Burt, writing in his notebook. "How would you get statistics on nature or anything else if somebody didn't do the basic counting?"

"Who gives a hoot about owls?" insisted Sheila.

"Everyone should. If it weren't for owls, our farms would be overrun with mice and our crops would be eaten," explained Burt. "But even if you don't care about owls, this kind of basic counting has to be done for things you **do** care about. For instance, how would you like it if manufacturers stopped producing blue jeans because they weren't able to get statistics on the cotton crop?"

"You mean if everyone was like me?" laughed Sheila. "And no one was willing to count the cotton crop?"

"Exactly," replied Burt. "I think owls are a lot more interesting than cotton, but in either case, without gathering facts there wouldn't be any statistics."

2 **Question:** How do we know owls eat mice?

If You Can't Count It All, Try the Handy Sample

To get practice for becoming President of the United States, Pam thought she'd run for student government president. But she was nervous about her chances and whether she was doing enough campaigning.

"Take an opinion poll," said Tom.

"You think I have time to chase a thousand kids all over town to see how they're going to vote?" asked Pam.

Tom explained that presidential candidates didn't track down every single one of the 100 million eligible voters either. Instead, to save time and money, their pollsters took a **sample** of opinion. They only counted a segment of voters selected at random, but this sample was enough to give them a good idea of the total vote.

"The trickiest part," Tom went on, "is deciding how small your sample can be and still be accurate. If you ask 12 kids, and nine say they will vote for you, that should indicate you're going to get ¾, or 75%, of the total vote. But 12 out of 1,000 kids is only 1.2% of your voters. A sample of 50 kids, or 5% of your voters, might show a little more or a little less than 75% of the vote. The bigger your sample, the less your chance of making a mistake in the forecast."

3 **Question:** Do opinion polls show that Americans believe in the energy crisis?

Experiments Replace Nature

"Burt had an easy job," Brian complained to Carol. "To count those owl nests he had only to gather facts that nature provided. Look what I'm stuck with: figuring out which of these five brands of bicycle tires is best."

"Design an experiment to test the different tires," said Carol. "And nothing is better than testing them in use."

In no time Carol had helped Brian round up five kids who agreed to test the tires on their bikes. They were to go biking for three hours and then return so Brian could check the tires.

Three hours later four of the kids were back. Pete's tires, brand Macho, looked as good as new. The tires on three of the other bikes showed some wear. Then Karen arrived, pushing her bike because both her brand Mirabelle tires were flat.

Brian declared Macho tires the best brand, Mirabelle as the worst, and the remaining as middling, not too good, but not too bad.

4 **Question:** What do these add up to: Stunt riding, carrying passengers, carrying things in your hand, riding in rain or snow or on ice, riding too fast, zigzagging in and out of traffic, hitching a ride on trucks, not stopping to look for pedestrians or oncoming traffic?

Comparisons Are Tricky

"What a waste of time!" Karen told Brian. "Not only did I have to push my bike the last two miles, but your experiment stank from the beginning."

"Pardon me?" said Brian, trying to keep cool.

"Your results are worthless," said Karen. "First of all, Pete has a new, expensive bike. The other three kids have bikes that are old but okay. But my bike is such a wreck the wheels wobble. On top of that, Pete told me he only rode slowly back and forth on his block and didn't do much mileage. The other three kids took the gravel road up to the mountain, so their tires **had** to look worse. And I not only went along that trail in the woods that's full of rocks, but I went like a demon and did the most mileage. And you've got the nerve to compare all this!"

"You're right," said Brian. "I'll have to do the experiment over, making sure to test under equal conditions: Bikes and tires in the same condition before the start; the same route and the same mileage for everyone. We could make it foolproof if we rotated bikes and bikers—since no two kids ride a bike the same way."

5 **Question:** If a car has to brake for your bike, how fast can it stop?

3 PRESENTING THE FACTS

"Show Me! Tell Me!"

Gathering facts or taking samples isn't enough. To make your statistics useful you have to present them to others, in graphics (charts, diagrams, pictures) or in words.

Communicating with graphics dates back to prehistoric people who engraved on stone stick-figures of the animals they hunted, or who recorded their own presence in a cave by having all the adults and children make hand prints on the walls. An ancient Chinese proverb says: "One picture is worth more than 10,000 words." Indeed, the more facts you've gathered, the bigger the need to condense them in one way or another.

The type of graphics used in statistics boils down

mostly to charts—made of circles, bars, or lines. A circle gives you the pie chart: it looks like a sliced pizza. The bar chart is like a xylophone: a series of bars of different lengths. Lines can be slanted up to show increases, or slanted down to show decreases, or made as jagged as lightning to show ups and downs—and can result in the fancy charts called frequency polygons.

Charts, however, can limit you, because you have to **show** them to someone, and who wants to carry charts around all the time? Sometimes it's more useful to sum up complicated information in one word or two. For these **numerical** descriptions, statistics has invented its own language, with words such as the "mean," the "median," the "mode," the "range." These will be explained after the charts.

6 **Question:** What has statistics been unable to invent?

As Anyone Can See: The Pie Chart

Whenever David and his brothers decided to look at televison they fought about which programs to watch. Finally, David thought that the only fair way was to keep a record of how often they ended up watching **his** favorite programs, or Clarence's, or little Timmy's. At the end of two weeks they had watched 70 hours of televison and the figures showed:

Favorite Programs	Timmy	David	Clarence
hours	3½	17½	49
percentage	5%	25%	70%

Because Timmy was too young to know about percentages, David decided to put his statistical results into a picture form that Timmy could understand. He chose the pie chart, one of the most common methods for showing a comparison of percentages. Of course, if David had had a hundred brothers and sisters sharing the television set, cutting the pie into so many slices would have made the chart a mess. But for only him, Clarence, and Timmy it was perfect:

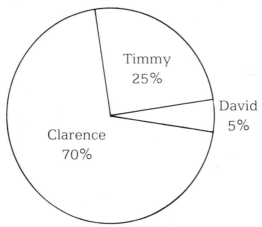

Even tiny Timmy could see who the television hog was: Clarence. And, from then on, David and Timmy were determined to use their pie chart power for equal television time.

7 **Question:** Even if David and his brothers share time equally, who's the hidden winner?

Bar Chart: The Mighty Rectangle

Although Ted was making money selling second hand books, some months he didn't seem to sell many and other months he sold a lot. He thought he could demonstrate the situation better if he put his sales figures into graphic form. He wasn't interested in percentages, so he didn't use a pie chart. Instead he used a common way of displaying economic data, the bar chart:

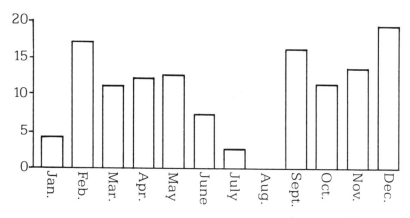

Putting his monthly figures this way, he could see at a glance that he made the most money over the Christmas holiday and when new school terms began. And he made almost no money when the kids were on summer vacation. "Probably they're too busy playing outdoors to read," he thought. "Well! I might as well close up my business in summer and take a vacation too!"

8 **Question:** In what other ways could Ted have made a bar chart?

A Matter of Speed:
The Frequency Polygon

Jean's dream was to win a gold medal swimming at the Olympics. But to become a champion, she had to keep careful notes on her speed to see whether she was improving. The record for the 100-meter free style was 0:55.65. Her notes for 25 trials looked like this:

0:55.65	0:55.60	0:55.62	0:55.59	0:55.55
0:55.75	0:55.65	0:55.61	0:55.54	0:55.60
0:55.69	0:55.70	0:55.60	0:55.65	0:55.61
0:55.65	0:55.71	0:55.60	0:55.65	0:55.65
0:55.69	0:55.70	0:55.69	0:55.60	0:55.65

But done this way, Jean couldn't tell at a glance just how she was doing. To condense all these data more visually she decided to construct a frequency polygon, a chart that would show how many times she had attained various speeds. She divided her speeds into five catagories, noting next to each how frequently her speed fell into each category:

much slower than the record	1 time
slower than the record	6 times
equalled the record	7 "
faster than the record	9 "
much faster than the record	2 "
	25 times

To construct the frequency polygon, she put her speeds along the horizontal line and her frequencies, the number of times she made each speed, along the vertical line.

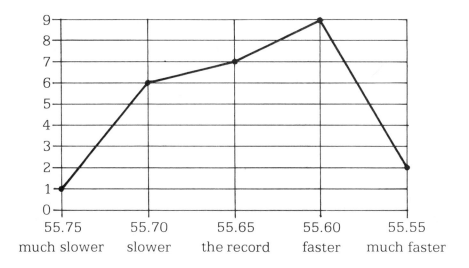

| 55.75 | 55.70 | 55.65 | 55.60 | 55.55 |
| much slower | slower | the record | faster | much faster |

9 **Question:** Do whales swim fast?

4 STATISTICAL LANGUAGE

The Mean

About to play the first football game of the season, Earl, the quarterback, needed a quick way to see how his record compared with that of the other team's quarterback. During the first quarter, Earl tried to pass 8 times, and he wrote down the yardage he gained each time:

5   0   0   10   6   8   0   15   yards

Adding these up, he found he had gained 44 yards. His teammates were dejected because the opposing quarterback, who had gained 48 yards in 12 attempts, seemed to be doing better.

"Cheer up," Earl said. "I have a better **mean** ." He explained that the mean, or the arithmetical average, is the **sum** of the measurements divided by the **number** of measurements. The sum of his yardage was only 44, but when divided by 8 attempts it gave him a mean of 5.5 yards per passing attempt. His opponent's total of 48 yards, divided by his 12 attempts, meant a mean of only 4 yards per attempt. By this widely used method of reducing a lot of statistical data to a few easily compared numbers, Earl tried to convince his teammates he was going to win.

"But the other guy is throwing more passes," one of his players objected.

"Aw, his arm will give out," promised Earl.

Sure enough, by the end of the game Earl had gained a total of 192.5 yards in 35 attempts, and his opponent had only 152 yards in 38 attempts!

10 **Question:** Is the mean noise level of schoolchildren mean on their teachers' ears?

The Median

To convince his parents that 25 cents a week wasn't enough, Paul thought it would be smart to tell them what allowances some of his friends received. One of his friends got 50 cents, two 75 cents each, and a fourth got $1 a week. But then he asked Howard, who was rich, and Howard said, ''I get $30 a week.''

''Rats,'' Paul said to Mary. ''That averages out to $6.60 a week. $6.60 is the mean. My parents will never give me that much. Besides, it's not fair to ask for so much, because Howard is an exception.''

''That's the trouble with mean,'' said Mary. ''If you're comparing just a few figures and one is very different from the others the mathematical average comes out screwy. Means are most truthful when you have a lot of comparisons—if you had asked 20,000 kids, Howard's $30 would have evened out. But since you can't take such an extensive survey, don't use the mean, use the **median** instead.''

Mary showed him how to get the median. First she lined the figures up in order of magnitude:

$.50 $.75 $.75 $1. $30.

and then she took the **middle** figure, or $.75. That way, she explained, Paul didn't have to count the distorting figures that were either too high or too low.

''By the way,'' said Paul, ''How much allowance do **you** get?''

''A buck,'' said Mary, as Paul lined up the figures again:

$.50 $.75 $.75 $1. $1. $30.

"Hey, now it doesn't work," said Paul. "There's no middle number."

"I forgot to tell you," said Mary, "you take the middle number only when you have an **odd** number of measurements. For an **even** number of measurements you have to take the two middle figures and average them."

Adding $1 and $.75, Paul got $1.75; dividing by two he got $.87½. "Fabulous," laughed Paul, "I'll tell my parents the median is 87½ cents and they'll go bananas trying to break a penny in two every week!"

11 **Question:** Which five U.S. cities have the highest median family income?

The Mode

Al was hoping to make money by selling cookies. But although he used nuts and fruits and other delicious ingredients, people wouldn't even try his cookies. Everyone just said that they couldn't afford $7 for a pound of cookies, no matter how good the ingredients were.

"You went at it backwards," said Harry. "Before making cookies you should have found the **mode** . It's a statistical measurement that's not as common as the mean or median, but it's useful in special cases like yours. If you have a group of numbers, the mode is the measurement that occurs most often."

"What does that have to do with cookies?"

"The mode is often used in selling. Before making high-quality, expensive cookies, you should find out what people are willing to spend for a pound of cookies. Then you can make the cookies at a reasonable price."

"Wow!" said Al when he returned after questioning the neighbors. "I really have to thank you for that good tip. Was I off base! Look at the answers I got:

$2, $1.50, $2, $2.50, $2, $2, $1.50, $2, $2, $3, $2, $2."

"Most people said $2," said Harry, "so that's your mode."

"Sshhh," said Al, leafing through recipe books. "I've got to find yummy cookie recipes without such expensive ingredients so I can sell them for $2 a pound."

12 **Question:** Out of every $20 spent on food, how much does the average family spend on bakery products?

27

The Range Can Kill You

"How come you never swim in Shallow Pond?" Marvin asked Sam. "It's so pretty, and huge too."

"Are you kidding?" said Sam. "How do you think it got its name? It averages only two lousy feet deep. In some places it's only a couple of inches deep."

Still, thought Marvin, it would be perfect for his younger brother Rocky, who was just learning to swim. "Since you're three feet tall, and the pond is only two feet deep, it's absolutely safe," Marvin encouraged Rocky a little later as they plunged into Shallow Pond. "Isn't this great?"

But there was no answer. Marvin stood up to look around and couldn't see Rocky. Only a second before he had been right there. All of a sudden, Marvin too began sliding into a deep hole. A good swimmer, Marvin got out of the hole with a couple of strokes. Then he pulled the struggling Rocky to safe ground.

"I almost drowned my brother because of your rotten two feet," Marvin told Sam angrily next day. "How could you have been so wrong?"

"Gee, I'm sorry," said Sam. "How was I to know you were going to take Rocky? Besides, when I told you it was only a couple of inches deep in some places, didn't you realize that two feet was the **mean** depth so that some spots had to be deeper? Here, I'll show you." Sam took Marvin to a map of Shallow Pond that hung in the school hallway.

"There are several deep holes," said Marvin, examining the map. "The deepest seems to be 12 feet."

"Sure, and a lot of places with only a couple of inches of

water; here it's only ½-inch deep," said Sam. "And if you add up all the different depths you do get a mean of two feet."

"What an idiot I am," moaned Marvin. "Of course, the shallow parts would have to be balanced by deeper spots to average out to two feet."

"You should have thought of the **range** instead of the mean," agreed Sam. "To get the range you use the same set of measurements as the mean but the range is the difference between the largest and smallest measurement."

"So the range for Shallow Pond would be the difference between the 12-foot depth and the ½-inch depth: 11 feet 11½ inches," said Marvin. "Even the tallest kids from the basketball team could drown in a two-foot-average-deep pond!"

13 **Question:** If the average annual temperature in Oklahoma City is a balmy 60.1 degrees, isn't it always comfortable?

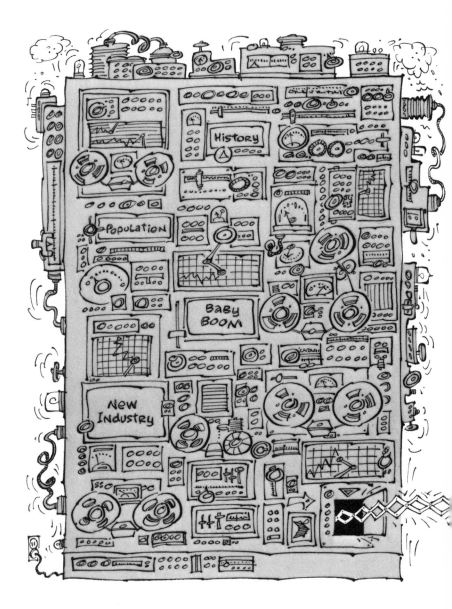

5 STATISTICS AS SCIENCE OF DECISION MAKING

Joan was worried. "How many orientation programs should I print next spring when I don't know how many newcomers will be coming to our high school?"

"Well, don't you have data about how many students came this spring, and the spring of last year, and the year before?" Nan asked.

"Of course," replied Joan. "165 this spring, 150 last spring, two years ago there were 137. But what good are these old figures?"

"They can help you quite a lot," said Nan. "Although the original job of statistics was the gathering and representing of facts, nowadays more and more people use these data to draw conclusions."

"Terrific," said Joan. "But how does that help me?"

"What you do is to use the enrollment figures you have, the ones for the last few years. Although these figures represent only a fraction of all the enrollment figures in our school's history, you can calculate that the increases from 137 to 150 is 13, and from 150 to 165 is 15, etc. So you might come up with a reasonable guess that next spring's enrollment will be about 14 more students."

"Great," said Joan. "Then I can print about 180 programs."

"Sure," said Nan, "but don't blame me if something vital
got left out of the calculations—like maybe the baby boom
is over and enrollments will be dropping rather than
increasing. Even the most careful conclusion can have a
flaw."

14 **Question:** Is it probable your hoped-for future income
will influence the amount of studying you do?

6 UNEXPECTED BY-PRODUCTS

A Fairy Tale Comes True

Centuries ago, according to a Persian fairy tale, three Princes of Serendip (an old name for Ceylon) went out looking for adventure and kept stumbling across unexpected treasures.

The word serendipity—coined after "Serendip"—therefore came to mean the aptitude for accidentally discovering valuable things while searching for something else.

But serendipity is also found in real life. Today in research it is known as **unexpected by-products**. A famous unexpected by-product is America: Columbus discovered it while searching for a route to Japan, China, and India (then known as "The Indies"). And often you hear of scientists whose studies in one field lead them to important discoveries in a different area. So it's not surprising that unexpected by-products play an interesting role in statistics too.

15 **Question:** What happens if you divide 9876543 by 8?

Pool and Lake Mix Well

When school was not in session, townspeople often jammed the sports facilities. Don decided that if he made and posted an hourly record of the number of people using each sports area, it would be possible to avoid the crowded times. His chart showed a surprise: two areas were used only half the time. The outdoor pool, jammed all summer, was drained in winter. The skating pond, crowded in winter, was used only for fishing in summer. Why not use both areas all year, he thought, and so prevent crowding at each? Why not make a swimming beach at the pond, and fill the pool in winter for ice skating? An unexpected reward or by-product of his study came when the school board voted to accept his suggestions and made Don student supervisor.

16 **Question:** How popular is swimming?

Swamp as a Bonus

At Town Meeting, Mildred volunteered to find the most common trees in the town's nature preserve and to label a nature trail. Mildred made a map of the preserve, and each day she'd walk through a different section of the forest, noting on her map how many beeches, maples, and other hardwoods she found. But after a while Mildred noticed that in one corner of the map her counts always included alders and willows. That meant moister earth. Sure enough, by exploring this area she discovered a beautiful swamp. "What a bonus!" Mildred thought. "After I label a hardwood trail, I'll build an interpretive boardwalk across the swamp!"

17 **Question:** What causes forest fires?

Snake Dance

To find out which books were needed for the school library, John was given the job of tabulating how many days of the year various types of books were taken out. John wasn't surprised that some popular books had been on loan 180 or more days a year. But he couldn't understand why **all** the books on reptiles were gone 365 days of the year! The withdrawal slips showed that Alice was the culprit. She had taken out the books, returned each one when it came due, and then checked it out again the same day. "Alice has an iguana, four frogs, five snakes, and three lizards," John explained to the librarian. "She wants to be a herpetologist."

"You're a pal," Alice told John later, dancing for joy. "Because of your study the librarian convinced my parents to **buy** me all the books I need."

18 **Question:** What is the life span of a snake?

Shoot Fast

The film club kept a record of how much each member spent on film and processing, and how much admission was collected at the club's movie showings. Most members found they spent so much on film that their shares of admissions covered only 35% to 50% of their expenses. But Brenda, who spent less than half of what the other members did, usually covered 100% of her expenses, and often she made a profit. "What's your secret?" the members asked Brenda. In discussion it became clear that Brenda spent twice as much time as they did preparing her scripts—so she wasted less film once she started shooting. Soon they were all spending a lot more time on their scripts and a lot less money on film!

19 **Question:** How far ahead should we plan for living in space?

7 LYING WITH STATISTICS

How Not to Be Conned

Like people in any profession, statisticians can err. A government employee going into a mountain community to count taxpayers might, despite diligent searching, find only 1,000 residents, whereas if he arrived offering free food stamps 1,400 people might come running out of the woods.

But this type of mistake is nothing compared to the mistakes, half truths, and out-and-out lies that occur when untrained people **innocently** misinterpret statistics or when other people **purposely** misrepresent statistics to sell a product or to prove a point.

To keep from being conned by impressive sounding figures, here are some questions to ask:

Who gathered the data and **why**? Was the research done by professional statisticians trying to find the truth, or by a group or individuals trying to prove a special point?

How was the data gathered? Was the sample large enough? Were safeguards taken to make sure the interviewers didn't cheat and that the interviewee didn't lie?

How are the results **presented**? Do you have all the facts you need to evaluate the statistics? Or are you given only percentages (not absolute figures) and incomplete charts?

20 **Question:** How many people lie?

38

Fathers and Children

For a discussion on family relationships, Bob and Leila were assigned to find out how much time fathers devoted to their children each evening.

"Let's not be dummies and ask the **kids**," said Bob. "Kids who get no attention might be ashamed and lie, so our results would be wrong."

Leila agreed they had to make their own observations. When they met 5 days later to compare results, Bob's figures showed the fathers he had observed had spent 2, 2½, 2½, 3, and 3 hours with their children. Leila's figures showed 10, 15, 15, 20, and 30 minutes.

"I don't get it; I went to a different house each night," Bob said, "and made careful notes on how long the father was around. How could our results be so different?"

"Because you were biased without realizing," said Leila, examining Bob's notes. "You counted **all** the time the father was in the house, even if he wasn't talking to his kids at all but watching TV, figuring accounts, taking a shower, or doing something else."

"So maybe I goofed," said Bob, reading Leila's notes. "But you're just as bad. All you counted was the time the father spoke **directly** to the kids. What about the time the whole family spent in discussion at dinner and things like that?"

"We both goofed," admitted Leila. "We'd better do the whole experiment over, maybe use a tape recorder so we can analyze every word more accurately and be able to have proof of what went on in case our analysis disagrees."

21 **Question:** How many families have both a mother and father?

Tarantulas vs. Poodles

"80% of Students Agree: Tarantulas Make Better Pets Than Poodles."

Betty scoffed when she read the headline Larry had used to advertise his insect business.

"You can't argue with statistics," said Larry.

"Baloney! The most common trick is to take a survey so small it's meaningless," said Betty. "Then, instead of revealing the puny sample size, to report the result as an impressive sounding percentage."

"Okay, so I did only ask five of my customers, but four said they agreed. That's 80%. What's wrong with that?"

"Not only is five kids out of the thousands in our city too small a sample," said Betty, "but by asking only your customers you've biased the sample by asking only people who are bug lovers."

"I was only trying to be professional," said Larry. "Big business ads never mention how many people were asked or who they were."

"Ads like that are for dopes," said Betty. "I'm going to buy a poodle."

22 **Question:** In England, in 1603, a schoolboy was whipped for refusing to smoke. Why were his teachers misled into thinking smoking would be good for him?

Who Likes to Babysit?

Jim and Ann did a study to find out if businesses in their neighborhood would discriminate against women when hiring for jobs. After they had each interviewed 35 prospective employees, here were the results:

	Jim's replies preferred		Ann's replies preferred	
	boy	girl	boy	girl
babysitters	0	35	7	28
gas station assnt	35	0	29	6
office helper	10	25	17	18
sales clerk	8	27	17	18
lawn mower	35	0	15	20

"How could the results be so different?" asked Jim. "What did we do wrong?"

"We didn't do anything wrong," said Ann. "It was the people we interviewed. A lot of them lied to me because they were too ashamed to admit they discriminate against women. The answers you got were a lot more honest because people didn't feel they had to hide their prejudice to a boy."

"Good thing we did this together," said Jim. "If two boys or two girls had done the survey, there wouldn't have been any discrepancy in the replies. We could have had the wrong results and never known it!"

23 **Question:** Might your sex cost you money in the job market?

Be an Optimist and Win

"Can you believe it?" Louise asked her brother Ben. "Only six kids out of 24 in my class are willing to help plant the vest pocket garden? That's not enough to get the work done. How am I going to convince more to join in?"

"Try presenting your figures differently," suggested Ben. "An optimist might say, 'Isn't it wonderful over three-quarters of our homes can afford television,' while a pessimist might say, 'Isn't it awful almost a quarter of our homes have no television.' In each case they are using the same numerical facts to bolster their arguments. So instead of saying that only six kids are going to help, say that a full quarter of the class has already signed up."

"You're a genius," said Louise later. "Once the kids heard that a full quarter of the class had signed up it had a psychological effect, the rest didn't want to be left out. So six more signed, and then five more volunteered."

24 **Question:** Which states have the most city and county parks?

Crazy Competitors

"I'm going to buy Crazy ski bindings," George told Ellen when he met her in the sports shop. "See how honest this company is—they give you full information on their test. Remember, in an emergency, your ski binding should loosen up, so that your foot can separate from the ski. Look," and he showed Ellen the ad he had clipped:

	kids	days	bindings loosened	bindings held firm
crazy ski binding	10	14	9	1
competitor's ski binding	10	14	2	8

"I don't trust that ad," said Ellen. "That survey was probably done by a research laboratory to **prove** Crazy bindings are better. It's only based on 20 kids and it doesn't say how many times they skied. Maybe in that test the kids in Crazy bindings went down the hill only once a day and the kids in Competitor's bindings went down ten times a day for 14 days. I'm going to rely on this survey done by a consumer group that had nothing to gain by the result, and which says that Competitor bindings are better."

Later, on the ski slope, Ellen saw the ski patrol lifting George onto a stretcher. George was screaming, "Wait till I find that research lab! I'm going to make **them** all ski 14 days on Crazy bindings."

25 **Question:** Why are skiing records frequently broken?

Catch as Catch Can

The new boy walking along the fishing pier stopped to watch Terry, Maria, and Bill. "Say, about how many fish do you catch a day?" he asked.

"About a dozen," said Terry, after thinking real hard.

"Seven or eight," said Maria.

"Five," said Bill.

The new boy laughed. "Two of you have to be wrong."

"No," said Terry. "We've all spoken the truth. It depends on whether you are speaking of the mean, the median, or the mode. We've got a dozen kids here and we write up our daily catches on the blackboard on the pier, see?" He pointed:

Albert	40 fish	Madeline	5 fish
Bobby	20 fish	Sean	5 fish
Maude	20 fish	Neil	5 fish
Debbie	10 fish	Frank	5 fish
Ed	10 fish	Edna	5 fish
Terry	10 fish	Jackie	5 fish

"Bill, who is too young to figure, knows that most kids catch about 5 fish a day, so he is right, he used the mode. Maria, knowing some kids catch more, thought it wasn't fair to say only 5, so she figured the median (7.5) was a more truthful answer. And me, I just averaged it and arrived at about a dozen."

"We're not only good fishermen," said Maria. "We're great statisticians too!"

26 **Question:** What are the causes of fish kills?

The Blown-up Chicken

To impress customers with how satisfied people were with his chickens, Alex wanted to display a chart of how his chicken-raising business had doubled in a year. First he tried a bar chart that looked like this:

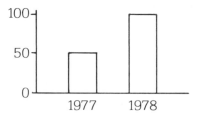

"It would look more dramatic," said his assistant Ruth, "if we took out the zero and showed only **half** the chart."

"Great idea," said Alex, as he cut off the bottom of the chart.

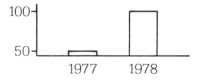

"It might look even more impressive," said Ruth thoughtfully, "if you used a picture instead of just bars."

"Terrific!" said Alex, as he began to draw. "Funny, even though one chicken is only double in height it looks a lot bigger."

1977   1978

"That's because," said Ruth, "by making the chicken twice as high you're also making it more than twice as wide. It's really cheating a bit, I think."

"Businesses do it all the time," said Alex, taking up the final version.

27 **Question:** How much chicken do we eat in the United States?

Sky and Skirts

''I'm going to be an airline pilot,'' said Eva.

''Girls aren't any good at that,'' said Scott. ''Less than one percent of the airline pilots are women.''

''You're stupid,'' said Eva. ''That's a common mistake in statistics, to draw the wrong inference from the facts. It may be true that less than one percent of the airline pilots are women, but that isn't because they're no good at it. It's because airline companies wouldn't hire them.''

''Maybe you're right,'' said Scott. ''It's easy to draw the wrong conclusion from figures.''

''Grandpa told me I shouldn't be a pilot,'' continued Eva, ''because airline fatalities are increasing. He showed me figures that only 551 passengers were killed in 1950, whereas in 1973, 824 passengers were killed. I looked it up later and found that he had drawn the wrong conclusion: there were more people killed only because so many more passengers were flying. When I checked the fatality rate per 100 million passenger miles flown, I found the fatality rate was a whopping 3.15 in 1950 and had dropped to .26 in 1973! But I didn't tell grandpa so as not to hurt his feelings.''

28 **Question:** A case of false conclusion occurred in a study of sugar cane workers. It was found that those ill with snail fever did more work than healthy cutters! What could have been wrong?

Up Against the Tent

"A free two-week camping trip!" Walter and Jane yelled as they rounded up 19 of their friends. "The Waterproof Tent Company is paying all expenses. Walter and 9 kids will test Waterproof tents in Death Valley. Jane and 9 kids will test the competitor's tents in Olympic National Park!"

At the end of two wonderful weeks, Walter and his group agreed that the Waterproof tents held up excellently, while Jane and her group all agreed that the competitor's tents were terrible. But when the Waterproof Tent Company ad came out, Jane ran to Walter. "Look at this dirty trick. The ad doesn't mention that my group camped in Olympic National Park's rain forest where the rainfall is 130 inches a year, and where it poured and poured."

"And it doesn't mention that my group camped in Death Valley," agreed Walter, "where the annual precipitation is only 2 inches and it never rained a drop."

"We didn't lie," said Jane. "And everything the ad says is true."

"It's what the ad **doesn't** say!" agreed Walter. "By leaving out an important fact about the test, this ad is a plain lie."

29 **Question:** Which of our National Parks is the most popular?

CONCLUSION:
NO LIMITS, NO CURE-ALL

Even if statistics doesn't fascinate you, it is here to stay. And it is growing so much that it seems to have no limits. Statistics has now invaded every field: insurance companies figuring your chances of living, archaeologists deciding where to dig, shoppers looking for the best buy. The list of uses just grows and grows.

Computers have had a big impact. When people had to figure with pencils or mechanical calculators, it was time-consuming and costly to compile statistics. Sophisticated computers are now giving us faster results at lower prices.

Like the computer, statistics itself is only an instrument. Properly used, statistics describes certain facts or helps you make decisions through inference. Once you misuse the instrument, or lie with it, the results will be deceptive or misleading. Misuse is not always intentional. Often surveys or experiments are poorly planned to begin with, and no later amount of juggling or goodwill can straighten out the initial mistake. Numbers are only as good as the people using them. So people who say they don't believe in statistics or who say you can prove anything with statistics are

mixing up the instrument with human error (innocent or intentional).

Statistics is no panacea, it can't solve all the problems on earth. Yet if statistics put you to sleep, you may be reading the wrong statistics. That most spiders have 8 eyes might bore an alligator lover; that alligators grow as long as 18 feet might bore a weatherman; that lightning strikes the earth 8 million times a day might bore you. But if a subject interests you, chances are its statistics will too.

ANSWERS

1. They don't understand statistics. They think that if they've lost 10 times—or 20 times or 30 times—they finally **have** to win. But mathematics doesn't care how many times you try. The chances of a penny coming up heads or tails are exactly the same on the first toss as on the 18th, 20th, 30th, or more tosses. And besides, if you consider **chances,** you're moving from statistics into probability.

2. One count of barn owl pellets showed 225 meadow mice, 179 house mice, 20 rats, 20 shrews, 1 mole, and 1 sparrow had been eaten.

3. A public opinion poll conducted by CBS News and **The New York Times** in 1977 showed that 38% of Americans thought the energy problem was real, 49% thought it was not real, and 13% weren't sure.

4. At least 30,000 bike accidents a year.

5. The average stopping distance for passenger cars on a dry road is:

at 20mph:	3 car lengths
30mph:	6 car lengths
40mph:	9 car lengths
50mph:	16 car lengths
60mph:	23 car lengths
70mph:	33 car lengths

6. The perfect way to present facts.

7. The advertisers. In 1976, 19.7% of all advertising in the U.S. was on television.

8. He could have made the bars out of books or coins:

9. Speeds of the great whales vary. The humpback, which averages 45 feet in length, cruises at only 2-5 mph and has a top speed of only 9-10 mph, while the little 27-foot minke easily does 16-21 mph. Fastest of all are probably the 68-82-foot fin and the 70-100-foot blue whales, which cruise at 11 mph or more and have been recorded swimming at 23 mph for more than 10 minutes. The 44-foot sei whale usually cruises at 5-6 mph and can go over 20 mph, but scientists doubt some reports that it has been seen swimming 40 mph.

10. A comfortable sound level is 60 decibels and the human ear begins to feel pain over 120 decibels. One study showed that elementary school children's shrieks have a mean level of 114 decibels, and one child's screech reached 122 decibels.

11. a.   Honolulu, Hawaii
    b.   San Jose, California
    c.   Seattle, Washington
    d.   Indianapolis, Indiana
    e.   St. Paul, Minnesota

12. 76 cents.

13. No. It has gotten as hot as 113 degrees and as cold as -17 degrees, so the range is a whopping 130 degrees.

14. Yes. In 1976, the median income of people who had finished:

	men	women
elementary school	$10,600	$5,691
high school	$13,542	$7,777
college	$19,658	$13,138

15. You get 1234567.

16. The U.S. Bureau of Outdoor Recreation estimates that 52% of the population swims (24% fishes, and less than 5% ice skates).

17. The U.S. Forest Service estimated in 1974 that 31.4% of forest fires were deliberate (arson), 23% were started accidentally from burning debris, 13.4% were started by careless smokers, 12.6% by lightning, 5.7% by machinery, 4.3% by campfires, and 12.1% by miscellaneous causes.

18. Small species may live 10 to 18 years, and large snakes more than 20 years. One anaconda kept in captivity lived 29 years.

19. Prof. Gerard K. O'Neill of Princeton University thinks we should have started already. Since 1969, he has been planning ahead for the problems involved in establishing colonies in space—including space-shuttles, using construction materials from the asteroids, and social problems—even though these colonies probably wouldn't be set up until after the year 2000.

20. Nobody knows for sure. No agency has collected

reliable statistics on this. Some researchers think everybody lies.

21. A few children who smoked cigarettes happened to avoid the plague, so teachers concluded that tobacco prevented the plague! They had used too small a sample.

22. In 1976, the U.S. Bureau of Census reported that 69.8% of American families were headed by a married couple. The remaining families were headed by a mother or father who was divorced,separated, widowed, or single.

23. Yes, and this is not a lie! In 1970, the median earnings for men in professional and technical jobs was $10,735 and for women $6,034. Male managers and administrators made $11,277 and females made $5,495. Salesmen made $8,451, saleswomen made $2,238.

24. California, with 2,371 parks covering 130,600 acres. Illinois has 1,354 parks but they cover only 38,500 acres, while the 1,109 parks in Texas cover 73,500 acres.

25. Expert ski racers in the 19th century were lucky to do 80-90 mph in bulky clothing and heavy skis. Today, speeds are over 220 mph—because of skin-tight clothing, equipment designed to lessen drag, better skis, and scientifically determined aerodynamic skiing positions.

26. In 1973, the U.S. Environmental Protection Agency reported that 37,814,000 fish were killed by water pollution. Of these 10,385,000 deaths resulted from poor sewerage and waste disposal by cities; insecticides and poisons used in agriculture, and chemicals and wastes from industrial operations were other big killers.

27. On an average, Americans eat 41.4 pounds of chicken a year (and 287 eggs).

28. Further study showed that the sick men were also the most experienced. New workers were not yet experienced, and had not been so exposed to snail fever.

29. In 1976, Great Smoky Mountains had 11.4 million visitors. Hot Springs had 4.7 million; more than 3 million people went to Grand Teton and Grand Canyon, and more than 2 million to Acadia, Rocky Mountain, Yosemite, Shenandoah, Olympic, and Yellowstone.

INDEX